Fire Fighters
to the Rescue

By Jack C. Harris
Illustrated by Pat Stewart

Editorial support and review by the National Fire Protection Association,
Quincy, Massachusetts, sponsors of the *Learn Not to Burn*® *Curriculum.*

A GOLDEN BOOK • NEW YORK
Western Publishing Company, Inc., Racine, Wisconsin 53404

Tommy knew something was wrong. He and his sister, Judy, had been playing in their backyard, waiting for their mother to come home from shopping. Tommy stopped and wrinkled up his nose.

"Do you smell something funny?" he asked.

"Yes," said Judy, "it smells like smoke. It could be a *fire*!"

Quickly they ran to their front yard. "Look," yelled Tommy as he pointed to the house across the street, "there's smoke coming from Mrs. Elly's house!"

"What should we do?" Judy screamed.

Tommy didn't say anything. He rushed inside his house and ran to the phone. Next to the phone, in an easy-to-see place, Tommy's dad had put a list of important phone numbers. One of them was the local fire department. Tommy dialed the number.

"Hurry," said Judy.

Before the phone on the other end had a chance to ring a second time, a man answered.

Quickly and calmly Tommy told the man Mrs. Elly's address and his own name and phone number.

"You've done a great job," the man said. "Now wait outside so you can direct the fire fighters when they arrive!"

"Yes, sir," said Tommy. He hung up the phone and ran outside. Soon he heard the distant wailing of sirens.

A police car came around the corner, its siren screeching. Then a huge fire engine came from the other direction. Several fire fighters jumped out of the engine's big cab.

Judy pointed to the house. "Look," she cried, "Mrs. Elly is trapped."

"Please keep back so we can get her down safely and put out the fire," a fire fighter told Tommy and Judy and several neighbors who had gathered. His voice sounded strange from behind his big breathing mask.

The fire fighters raised a ladder to Mrs. Elly's window and carried her to safety.

"Your call probably saved Mrs. Elly's life," Fire Chief Richards told Tommy and Judy. "There was very little fire damage, too. How would you like to have a personal tour of the firehouse tomorrow as my guests?"

"Wow," said Tommy, "can we really?"

"Sure," the chief said, "if it's OK with your parents."

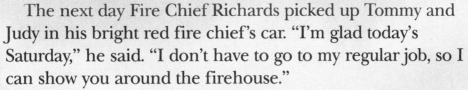

The next day Fire Chief Richards picked up Tommy and Judy in his bright red fire chief's car. "I'm glad today's Saturday," he said. "I don't have to go to my regular job, so I can show you around the firehouse."

"Aren't you a fire fighter all the time?" asked Judy.

"Oh, no," the chief explained. "In our town all of the fire fighters are volunteers. In big cities there are full-time fire fighters. Some communities have both volunteers and full-time career fire fighters."

Soon the chief's car pulled into the fire station parking lot. The station's large doors were open, and Tommy and Judy could see several big bright fire engines and trucks.

"Wow," said Tommy, "there are so many different ones."

"That's right," said Chief Richards. "We have special fire trucks for house fires, ones for grass or field fires, and ones for fires in bigger buildings."

PONDVIEW   ENGI

"I'll bet the one with the big ladder is for the big buildings," Judy said, looking at a fire engine with a long aluminum ladder folded on the back.

"That's exactly right," the chief said.

Inside the firehouse, Chief Richards introduced Tommy and Judy to one of the fire station's radio dispatchers.

"When you called our number yesterday, Tommy," the dispatcher explained, "you were connected to the police station, since all of our fire fighters are volunteers. They called us because we were the closest fire station to Mrs. Elly's house. We radioed our volunteers, who all wear special receivers called pagers. No matter where the fire fighters are, we can use the pagers to tell them immediately that there is a fire or an emergency.

"But our jobs don't end with that," the dispatcher continued. "We have to prepare important information for the fire fighters to see as soon as they come in. We do that on our 'ready' sign. The flashing sign tells the fire fighters exactly what type of fire or emergency there is so they can prepare the right equipment and fire engines.

"On the chalkboard below the sign, we write the exact location of the fire. The drivers look for the place on the map as soon as they arrive."

BRUSH   SCUBA
AUTO   ASSISTANCE
STAND BY   STRUCTURE

"The fire fighters also check the location as they put on their coats, gloves, helmets, and other equipment," said Chief Richards. "Each station in our town covers about a ten-square-mile area. Sometimes, if there's a big fire, we help other fire stations. That's when we'd light up the word *assistance* on the ready sign.

"When everything is ready, the fire fighters climb into their seats near the front of the fire engine and buckle their seat belts. Not too long ago fire fighters rode by hanging on to the sides or the rear of a speeding fire truck. This was stopped because it was too dangerous.

"A radio operator in the truck keeps in constant touch with the police and the firehouse. The operator relays any new information about the fire to the crew so they can be ready as soon as they arrive."

"I know what happens next," Tommy said excitedly. "The driver turns on the siren, and everyone gets out of the way."

"That's true," the chief said, "but the fire fighters don't rely on the sound of the siren alone. The radio operator contacts the police in the area, and they make sure that traffic stops in all directions and the streets are clear.

"You'll remember," the chief reminded Judy and Tommy, "that yesterday the police arrived at Mrs. Elly's house before the fire fighters. They usually get to a fire first because they are always on patrol and on the alert for emergency calls. The police can radio the fire fighters who are on the way to the fire and tell them exactly what's happening at the scene.

"When a fire is out," the chief said, "I stay behind to make sure the area is safe. As soon as the fire fighters get back to the firehouse, they make certain that the engines are ready for the next fire.

"The fire fighters fill the engines' gas tanks and fold up the hoses. They also check all their equipment. Each fire fighter has to have gloves, boots, a helmet, an air tank, an oxygen mask, and a raincoat. It costs a lot of money to equip one fire fighter properly," the chief said.

Then Chief Richards handed Tommy and Judy special plastic fire helmets of their very own.

"Wow, thanks!" said Tommy and Judy.

A PROPERLY EQUIPPED FIRE FIGHTER

HELMET
OXYGEN MASK
AIR TANK
PIKE POLE
AX
GLOVES
RAINCOAT
BOOTS

"Can we be fire fighters when we grow up?" Tommy asked.

"Certainly," answered Chief Richards. "You have to be eighteen to begin the training. Our fire fighters go to classes two nights a week for twenty-five weeks. We train on a special outdoor range and in a classroom to learn how to fight all kinds of fires, how to rescue people, and how to save lives.

"The techniques for fighting fires are always being improved," Chief Richards continued. "Not only that, but fighting the different kinds of fires requires specialized training. For instance, the fire fighters who combat forest fires learn methods that wouldn't be used for house fires.

"The same is true with the harbor patrols who fight fires on the water. They have to be expert sailors as well as specialized fire fighters."

"This sure has been fun," said Tommy.

"I've enjoyed showing you how we fire fighters do our jobs," the chief said. "Why don't you look around the firehouse. Your parents will be picking you up soon, and—"

But Chief Richards didn't get to finish his sentence. Suddenly the fire siren blared. The dispatcher Tommy and Judy had met called out, "There's a house fire!"

The electric ready sign started flashing *structure*, and fire fighters began running in all directions.

Tommy and Judy stood out of the way and watched as the men and women of the fire station prepared to battle the fire. Each one knew exactly where to go and what to do.

The big fire engine roared out of the firehouse, its siren wailing. Tommy and Judy watched as the big truck headed toward the scene of the fire.

As they watched, Tommy and Judy realized that their fire fighters were just a few of the thousands of dedicated men and women who battled blazes in towns and cities in every state. Somewhere, every hour of the day, all over the country, there are fire fighters to the rescue!